MAC'S FIELD GUIDE™

TO GREAT SMOKY MOUNTAINS NATIONAL PARK

WILDFLOWERS

KEY—COMMON NAME
Scientific Name
Habitat

ART BY DAVID SAUSKOJUS

INDIAN PIPE
Monotropa uniflora
Height: 7"
Moist, rich woods

PAINTED TRILLIUM
Trillium undulatum
Height: 14"
Moist, rich woods, Dry woods

WHITE TRILLIUM
Trillium grandiflorum
Height: 13"
Moist, rich woods

TRAILING ARBUTUS
Epigaea repens
Height: 5"–6"
Dry woods, roadsides, rocky soils

SHARP-LOBED HEPATICA
Hepatica nobilis var. acuta
Height: 5"
Moist, rich woods

SQUAWROOT
Conopholis americana
Height: 7"
Dry woods

FRASER'S SEDGE
Cymophyllus fraserianus
Height: 10"
Moist, rich woods

MAY-APPLE
Podophyllum peltatum
Height: 15"
Moist, rich woods

BLOODROOT
Sanguinaria canadensis
Height: 9"
Moist, rich woods

WOOD-SORREL
Oxalis montana
Height: 4"
Moist, rich woods

SPRING-BEAUTY
Claytonia caroliniana
Height: 7"
Moist, rich woods

DUTCHMAN'S BRITCHES
Dicentra cucullaria
Height: 8"
Moist, rich woods

FRINGED PHACELIA
Phacelia fimbriata
Height: 6"
Moist, rich woods

RUGEL'S RAGWORT
Rugelia nudicaulis
Height: 16"
Moist, rich woods

RUE-ANEMONE
Thalictrum thalictroides
Height: 6"
Moist, rich woods, open woods

WHITE BANEBERRY
Actaea pachypoda
Height: 18"
Moist, rich woods

FOAM FLOWER
Tiarella cordifolia
Height: 9"
Moist, rich woods, streams

GALAX
Galax urceolata
Height: 18"
Dry woods

SWEET JOE-PYE-WEED
Eupatorium purpureum
Height: 78"
Open woods

PINK LADY'S SLIPPER
Cypripedium acaule
Height: 11"
Dry woods

WILD GERANIUM
Geranium maculatum
Height: 20"
Moist, rich woods, meadows, roadsides

CANADA GOLDENROD
Solidago canadensis
Height: 36"
Meadows, roadsides, clearings

FILMY ANGELICA
Angelica triquinata
Height: 36"
Meadows, open woods, streams

MOUNTAIN BUGBANE
Cimicifuga americana
Height: 48"
Moist, rich woods, open woods

BEE-BALM
Monarda didyma
Height: 36"
Moist, rich woods, moist soil, streams

NEW YORK IRONWEED
Vernonia noveboracensis
Height: 54"
Meadows, moist soil, streams

SHOWY ORCHIS
Galearis spectabilis
Height: 8"
Moist, rich woods

FIRE PINK
Silene virginica
Height: 18"
Rocky soil, open woods, roadsides

TROUT-LILY
Umbilicatum americanum
Height: 7"
Moist, rich woods

CONEFLOWER
Rudbeckia laciniata var. humilis
Height: 48"
Moist, rich woods, meadows, moist soil

YELLOW TRILLIUM
Trillium luteum
Height: 9"
Moist, rich woods

CRANE-FLY ORCHID
Tipularia discolor
Height: 12"
Moist, rich woods

WILD GINGER
Asarum canadense
Height: 8"
Moist, rich woods, rocky soils

LITTLE BROWN JUGS
Hexastylis arifolia
Height: 5"
Dry woods, moist, rich woods

CARDINAL FLOWER
Lobelia cardinalis
Height: 36"
Moist soil, streams

ORANGE JEWELWEED
Impatiens capensis
Height: 42"
Moist soil

YELLOW-FRINGED ORCHID
Platanthera ciliaris
Height: 18"
Dry woods, meadows

TURK'S CAP LILY
Lilium superbum
Height: 66"
Moist, rich woods, meadows, moist soil

THYME-LEAVED BLUETS
Houstonia serpyllifolia
Height: 4"
Moist, rich woods, roadsides, streams

CRESTED DWARF IRIS
Iris cristata
Height: 6"
Moist, rich woods, open woods

PURPLE PHACELIA
Phacelia bipinnatifida
Height: 18"
Moist, rich woods, rocky soils

JACK-IN-THE PULPIT
Arisaema triphyllum
Height: 24"
Moist, rich woods

MAC'S FIELD GUIDE™

TO GREAT SMOKY MOUNTAINS NATIONAL PARK

TREES

KEY—COMMON NAME
Scientific Name
Habitat

ART BY DAVID SAUSKOJUS

TABLE MOUNTAIN PINE
Pinus pungens
Height: 20'–60';
Needles: 1¼"–3"
Cones: 2"–3½"; Range: 3000'–4500'
Dry, rocky soils

RED SPRUCE
Picea rubens
Height: 50'–80';
Needles: ½"–⅝"
Cones: 1¼"–1 ⅝";
Range: 4000'–6000'
Rocky soils

FRASER FIR
Abies fraseri
Height: 30'–50';
Needles: ¼"–1½"
Cones: 1½"–2½"; Range: 4500'-6650'
Conifer forests

EASTERN HEMLOCK
Tsuga canadensis
Height: 60'–70'; Needles: ⅜"–⅝"
Cones: ½"–1"; Range: 3500'–5000'
Moist, dry, or rocky soils

FLAME AZALEA
Rhododendron calendulaceum
Height: 10'–20'
Leaves: 1"–3"
Range: 0'–5800'
Mixed forests

SILVERBELL
Halesia tetraptera
Height: 10'–30'
Leaves: 3"–7"; Range: 0'–5000'
Hardwood forests, streams

YELLOW BIRCH
Betula alleghaniensis
Height: 70'–100'; Leaves: 3"–5"
Fruit: ½"–⅝"; Range: 3500'–5000'
Moist, rich woods, hardwood forests, mixed forests

MOUNTAIN LAUREL
Kalmia latifolia
Height: 10'–35'; Leaves: 2"–5"
Range: 0'–5000'
Dry soils, mixed forests

ROSEBAY RHODODENDRON
Rhododendron maximum
Height: 10'–40'; Leaves: 4"–10"
Range: 0'–5000'
Moist soils, hardwood forests, streams

SOURWOOD
Oxydendrum arboreum
Height: 20'–50'; Leaves: 3"–7"
Fruit: ⅜"; Range: 0'–4500'
Moist soils, mixed forests

AMERICAN BEECH
Fagus grandifolia
Height: 60'–80'; Leaves: 2"–6"
Fruit: ½"–¾"; Range: 0'–5800'
Moist, or dry soils

UMBRELLA MAGNOLIA
Magnolia tripetala
Height: 10'–35'; Leaves: 10"–24"
Fruit: 2½"–4"; Range: 0'–2500'
Moist, rich woods, hardwood forests, streams

CATAWBA RHODODENDRON
Rhododendron catawbiense
Height: 10'–20'; Leaves: 3"–8"
Range: 1500'–6500'
Rocky soils, streams

SUGAR MAPLE
Acer saccharum
Height: 70'–100'; Leaves: 3"–7"
Fruit: 1"–1½"; Range: 0'–5000'
Moist soils

RED MAPLE
Acer rubrum
Height: 60'–90'; Leaves: 3"–8"
Fruit: ¾"–1"; Range: 0'–6000'
Moist soils, moist rich woods, hardwood forests, streams

SWEETGUM
Liquidambar styraciflua
Height: 50'–120'; Leaves: 3"–6"
Fruit: 1"–1½"; Range: 0'–2000'
Moist soils, mixed forests, streams

YELLOW BUCKEYE
Aesculus flava
Height: 70'–90'; Leaves: 4"–7"
Fruit: 2"–3"; Range: 1000'–6300'
Moist soils, hardwood forests

SYCAMORE
Platanus occidentalis
Height: 50'–130'; Leaves: 4"–8"
Fruit: ¾"–1½"; Range: 0'–3000'
Moist soils, mixed forests, streams

TULIPTREE
Liriodendron tulipifera
Height: 50'–120'; Leaves: 3"–6"
Range: 0'–4000'
Moist soils, moist, rich woods

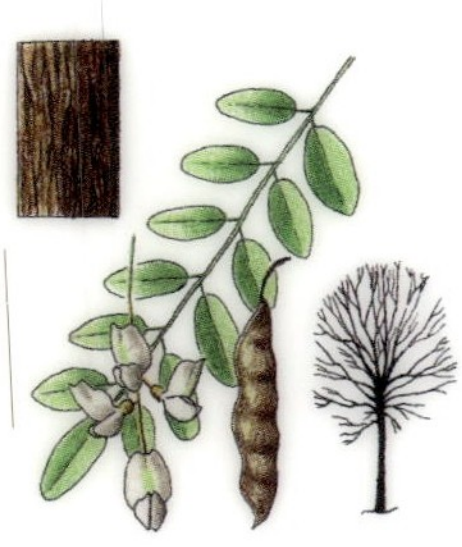

BLACK LOCUST
Robinia pseudoacacia
Height: 40'–80'; Leaves: 6"–12"
Fruit: 2"–6"; Range: 500'–5000'
Moist, dry, or rocky soils

50495
9 780898 866506

The Mountaineers, 1001 SW Klickitat Way, Suite 201,
Seattle, WA 98134; (206) 223-6303
ISBN 0-89886-650-2